YOUR KNOWLEDGE HAS VALUE

Sana Munir

The "Green's Function" Associated with One- and Two-Dimensional Problems

GRIN Publishing

Bibliographic information published by the German National Library:

The German National Library lists this publication in the National Bibliography; detailed bibliographic data are available on the Internet at http://dnb.dnb.de .

Imprint:

Copyright © 2015 GRIN Verlag GmbH
Print and binding: Books on Demand GmbH, Norderstedt Germany
ISBN: 978-3-656-95095-0

This book at GRIN:

http://www.grin.com/en/e-book/298668/the-green-s-function-associated-with-one-and-two-dimensional-problems

GRIN - Your knowledge has value

Since its foundation in 1998, GRIN has specialized in publishing academic texts by students, college teachers and other academics as e-book and printed book. The website www.grin.com is an ideal platform for presenting term papers, final papers, scientific essays, dissertations and specialist books.

Visit us on the internet:

http://www.grin.com/

http://www.facebook.com/grincom

http://www.twitter.com/grin_com

Green's Function Associated with one and two dimensional Problem

Sana munir

Department of Mathematics, GC University, Faisalabad.

In mathematics a green's function is type of function used to solve inhomogeneous differential equations subject to specific initial conditions or boundary conditions. Green's functions provide an important tool when we study the boundary value problem. They also have intrinsic value for a mathematician.

Also green's functions in general are distribution, not necessarily proper function. Green functions are also useful for solving wave equation, diffusion equation and in quantum mechanics, where the green's function of the Hamiltonian is a key concept, with important links to the concept of density of states. The green's function as used in physics is usually defined with the opposite sign that is

$$\ell\, G(x,s) = -\delta(x-s)$$

This definition does not change significantly any of the properties of the Green's function in heat conduction we know that the Greens' function represents that temperature at a field point due to a unit heat source applied at source point. In electro static the green's function stand for the displacement in the solid due to the application of unit point force.

In this project construction of green's function in one and two dimension has shown. There are more then one way of constructing greens' function (if it exist) but the result is always same. Due to this we can say that green's function for a given linear system is unique.

We start with the brief introduction of the Dirac delta or Dirac's delta function which is not strictly a function in real sense of functions.

Dirac Delta Function:-

    i. $\delta(t-a) = 0$ if $t \neq 0$

    ii. $\delta(t-a)dt = 1$

    iii. $\delta(t-a)f(t)dt = f(x)$

    iv. $\delta(t-a) = \delta(a-t)$

Green's Function Associated with one dimensional boundary value problem:

    Consider the following boundary value problem.

$$M[y] = F(x) \rightarrow (1) \; x_1 < x < x_2$$

$$B_1[y] = \propto \qquad B_2[y] = \beta$$

M is defined by

$$M = A_2(x)\frac{d^2}{dx^2} + A_1\frac{d}{dx} + A_0(x)$$

From(1)

$$A_2(x)\frac{d^2y}{dx^2} + A_1(x)\frac{dy}{dx} + A_0(x)y = F(x)$$

Divide by $A_2(x)$ we will get

$$\frac{d^2 y}{dx^2} + \frac{A_1(x)}{A_2(x)}\frac{dy}{dx} + \frac{A_0(x)}{A_2(x)}y = \frac{F(x)}{A_2(x)} \rightarrow \quad (2)$$

Let

$$P(x) = e^{\int \frac{A_1(x)}{A_2(x)}dx}$$

Multiply equation (2) by P(x)

$$P(x)\frac{d^2 y}{dx^2} + \frac{A_1(x)}{A_2(x)}P(x)\frac{dy}{dx} + \frac{A_0(x)}{A_2(x)}P(x)y = \frac{P(x)F(x)}{A_2(x)} \Rightarrow \frac{d}{dx}\left[P(x)\frac{dy}{dx}\right] + q(x)y = f(x)$$

$$\left[\frac{d}{dx}\left(P(x)\frac{d}{dx}\right) + q(x)\right]y = f(x)$$
$$\ell[y] = f(x)$$

Where

$$\ell = \frac{d}{dx}\left[P(x)\frac{d}{dx}\right] + q(x)$$

Consider the self-ad-joint boundary value problem

$$\left.\begin{array}{l} \ell[y] = f(x) \\ B_1[y] = \alpha, B_2[y] = \beta \end{array}\right\} \rightarrow (3)$$

Now split the boundary value problem into the following boundary value problems.

$$\left.\begin{array}{l} \ell[y] = 0 \\ B_1[y] = \alpha, B_2[y] = \beta \end{array}\right\} \rightarrow (4)$$

$$\ell[y] = f(x)$$
$$\left. B_1[y] = 0, B_2[y] = 0 \right\} \rightarrow (5)$$

The solution of problem (3) is written as

$$y = y_H + y_P$$

y_H is complementary function which satisfy the homogeneous differential equation

$$\ell y_H = 0$$

y_P is the particular solution to the inhomogeneous differential Equation

$$\ell y_P = +f(x)$$

The general solution of (4) is written as

$$y_H = c_1 y_1 + c_2 y_2 \rightarrow (6)$$

Here y_1 & y_2 are solution of homogeneous Equation $\ell[y] = 0$ *and* c_1, c_2 are arbitrary constant and they can be determined by applying the boundary condition.

Now consider the problem (5). Suppose that the solution of problem '5' can be expressed in the integral form as

$$y_P = -\int_{x_1}^{x_2} g(x,s)f(s)ds \rightarrow (7)$$

Where g(x,s) is the green's function which is to be defined later. The negative sign in (7) describe the physical interpretation. Apply the differential operator ℓ on both side of '7'

$$\ell[y_P] = \ell\left(-\int_{x_1}^{x_2} g(x,s)f(s)ds \right)$$

Since ℓ & the integral operator commute each other.

$$\ell[y_P] = -\int_{x_1}^{x_2} \ell(g)f(s)ds \rightarrow (i)$$

Also

$$-\int_{x_1}^{x_2} \delta(x-g)f(s)ds = f(x) \rightarrow (ii)$$

And also we have

$$\ell[y_P] = f(x) \rightarrow (iii)$$

From i, ii, and iii.

$$-\int_{x_1}^{x_2} \ell(g)f(s)ds = -\int_{x_1}^{x_2} \delta(x,s)f(s)ds$$

$$\Rightarrow -\int_{x_1}^{x_2} [\ell[g] + \delta(x-s)] f(s) ds = 0$$

Since $f(s)$ is arbitrary $f(s) \neq 0$

$\ell[g] + \delta(x-s) = 0$

$\ell[g] = -\delta(x-s) \rightarrow (8)$

$\delta(x-s)$ is dirac delta function.

To, determine the unique green's function condition (8) is not enough. We have to determined other condition also from homogeneous boundary condition.

$B_1[y] = 0$, $B_2[y] = 0$

$B_1[y_P] = 0$, $B_2[y_P] = 0$

$$B_1 \left[-\int_{x_1}^{x_2} g(x-s) f(s) ds \right] = 0$$

$$B_2 \left[-\int_{x_1}^{x_2} g(x-s) f(s) ds \right] = 0$$

$$\Rightarrow \int_{x_1}^{x_2} B_1[g] f(s) ds = 0$$

And

$$\int_{x_1}^{x_2} B_2[g] f(s) ds = 0$$

Since $f(s)$ Can be almost any function the above relations are satisfied only if.

$$B_1[g]=0 \quad , \quad B_2[g]=0 \rightarrow (9)$$

Hence the green's function we are looking for is solution of the following boundary problem

$$\ell[g]=-\delta(x-s) \rightarrow (10)$$

Related with boundary condition

$$B_1[g]=0 \quad , \quad B_2[g]=0$$

Where 's' is a fix value lies between x_1 and x_2 and. The above problem is similar to that given equation in (5) only the forcing function in (9) is delta function rather then arbitrary function f(x). This means that solving the problem for g is simpler then solving the corresponding problem 'y'.

And once the green's function has been determined for particular operator £ and set of boundary conditions it may be used for solving problem (5) for any umber of time where only the function f(x) changes from problem to problem. It is this feature of green's function that make it most useful I application.

The green's function g(x,s) associated with boundary value problem

$$\ell[y]=f(x)$$

$$B_1[y]=\alpha \quad , \quad B_2[y]=\beta$$

Where

$$\ell = \frac{d}{dx}\left[P(x)\frac{d}{dx}\right]+q(x)$$

Satisfying..................

(a) $\quad \ell[g] = -\delta(x-s) \qquad\qquad x_1 < x < x_2$

(b) $\quad B_1[g] = 0 \qquad , \qquad\qquad B_2[g] = 0$

(c) $\quad g(s^+,s) = g(s^-,s)$

(d) $\quad \dfrac{\partial g(s^+,s)}{\partial x} - \dfrac{\partial g(s^-,s)}{\partial x} = \dfrac{-1}{P(s)}$

The condition 'd' is called jumped discontinuity of green's function at x=5.

Based on above condition an exploit formula for the green's function can be designed. It is observed from the condition (a) if either x<s or $x \geq s$ then

$\ell[g] = 0$ By definition of Dirac Delta function if .... are solution of homogeneous differential equation $\ell[g] = 0$ Such that $B_1[Z_1] = 0 \quad B_2[Z_2] = 0$ so, from condition (a) and (b) the green's function has following form

$$g(x,s) = \begin{cases} u(s)Z_1(x)x < s \\ v(s)Z_2(x)x \geq s \end{cases}$$ where u and v are function to be determined imposing the condition c and d. the unknown function u & v must be chosen such that

$$v(s)Z_2(s) - u(s)Z_1(s) = 0$$

$$v(s)Z_2'(s) - u(s)Z_1'(s) = \frac{-1}{P(s)}$$

By solving the above two simultaneous equation we will get

$$v(s) = \frac{-Z_1(s)}{P(s)W(Z_1,Z_2)(s)}$$

$$W(Z_1,Z_2) = Z_1Z_2' - Z_2Z_1'$$

Since Z_1 and Z_2 are such solution we can write

$$P(s)W(Z_1,Z_2)(s) = P(s)W(Z_1,Z_2)(x)(s) = c$$

The green's function can be written as

$$g(x,s) = \begin{cases} \dfrac{-Z_1(x)Z_2(s)}{c}, & x_1 < x < s \\[2mm] \dfrac{-Z_1(x)Z_2(s)}{c}, & s \le x < x_2 \end{cases}$$

Green's function is symmetric in x and s.

i.e g(x,s) = g(s,x)

Green's function associated with two dimensional problem:

To, introduce green's function in two dimensions, we consider the PDE (poisson's equation which here represents static deflection of a rectangular membrane).

$$\frac{\partial^2 u}{\partial x^2} + \frac{\partial^2 u}{\partial y^2} = f(x, y) \rightarrow (11)$$

Here f(x,y) represent the external load per unit area, divided by T(tention in the membrane, which here has the dimension of force per length)

The B.CS. in this case are $u(0, y) = u(a, y) = 0$, $u(x,0) = u(x,b) = 0 \rightarrow (12)$

A concentrated force acting at a point (x',y') may be simulated by the two dimension delta

function $\quad \dfrac{F}{\tau}\delta(x - x')\delta(y - y')$

Let $G(x|x', y|y') = 0$ be the green's function associated with the problem

$$\frac{\partial^2 G}{\partial x^2} + \frac{\partial^2 G}{\partial y^2} = \frac{F}{\tau}\delta(x - x')\delta(y - y') \rightarrow (13)$$

G(0,y) = G(a,y) = 0, G(x,0) = G(x,b) = 0 → (14)

Then the solution of the problem(11) and (12) is given by

$$u(x, y) = \int\limits_0^a \int\limits_0^b G(x|x', y|y') f(x', y') dx' dy'$$

We first find a complete set of Eigen functions of the associated homogeneous Eigen value problem viz.

$$\left(\frac{\partial^2}{\partial x^2} + \frac{\partial^2}{\partial y^2} \right) u_\lambda(x, y) = \lambda u_\lambda(x, y) \text{ with the same B.CS as on } u(x,y).$$

We find $\quad \lambda_{mn} = -\left(\frac{m^2 \pi^2}{a^2} + \frac{n^2 \pi^2}{b^2} \right), m, n = 1,2,3,...$

and

$$u(x, y) \equiv u_{mn}(x, y) = \frac{2}{\sqrt{ab}} \sin \frac{m\pi x}{a} \sin \frac{n\pi y}{b}$$

which have been normalized to unity. Now the required green's function can be represented as $G(x|x'; y|y') = \frac{2}{\sqrt{ab}} \sum\limits_{m=1,n=1}^{\infty} A_{mn}(x', y') \sin \frac{m\pi x}{a} \sin \frac{n\pi y}{b}$

Now the required green's function can be represented as

$$G(x|x'; y|y') = \frac{2}{\sqrt{ab}} \sum\limits_{m=1,n=1}^{\infty} A_{mn}(x', y') \sin \frac{m\pi x}{a} \sin \frac{n\pi y}{b}$$

Substituting this into (13)

$$\sum\limits_{m,n}^{\infty} \left[-\frac{m^2 \pi^2}{a} - \frac{n^2 \pi^2}{b} \right] A_{mn}(x', y') \sin \frac{m\pi x}{a} \sin \frac{n\pi y}{b}$$

$$= \frac{F}{\tau} \delta(x - x') \delta(y - y')$$

Multiplying by $\sin(m'\pi y/a) \sin(n'\pi y/b)$ and integrating w.r.t x and y we have.

$$-\frac{2}{\sqrt{ab}}\left(\frac{m^2\pi^2}{a^2}+\frac{n^2\pi^2}{b^2}\right)A_{mn}\frac{a}{2}\cdot\frac{b}{2}=\frac{F}{\tau}\sin\frac{m\pi x'}{a}\sin\frac{n\pi y'}{b}$$

Or

$$-\left(\frac{m^2\pi^2}{a^2}+\frac{n^2\pi^2}{b^2}\right)A_{mn}(x',y')=-\frac{2}{\sqrt{ab}}\frac{F}{\tau}\sin\frac{m\pi x'}{a}\sin\frac{n\pi y'}{b}$$

Hence on substitution

$$G\left(x|x';y|y'\right)=\frac{-4}{\sqrt{ab}}\sum_{m,n}^{\infty}\frac{a^2b^2}{b^2m^2\pi^2+a^2n^2\pi^2}\frac{F}{\tau}\sin\frac{m\pi x'}{a}\sin\frac{n\pi y'}{b}\sin\frac{m\pi x}{a}\sin\frac{n\pi y}{b}$$

$$=-\frac{4abF}{\pi^2\tau}\sum_{m,n=1}^{\infty}\frac{1}{b^2m^2+a^2n^2}\sin\frac{m\pi x'}{a}\sin\frac{n\pi y'}{b}\sin\frac{m\pi x}{a}\sin\frac{n\pi y}{b}$$

References

- Boas, M.L. (1983). Mathematical Method in physical Sciences. (2nd Edition). John Wiley and sons, New York

- Khalid Latif Mir. (1997). Problem and Methods in Mathematical Physics and Applied Mathematics. Markazi Kutab Khana, Urdu Bazar, Lahore.

- Barton, G (1989). Element of Green's function and Propagation. Oxford, UK